Kenwood

The Iveagh Bequest

Laura Houliston and Susan Jenkins

CONTENTS

Facing page: *The south front of Kenwood, photographed by John Gay in the 1960s*

Left: *View of the north front, which has been redecorated with a sanded paint to resemble sandstone, reflecting its appearance in the 1790s*
Below: *Medallion of Robert Adam by James Tassie, 1792*
Bottom: *Engraving of the south front of Kenwood from* The Works in Architecture of Robert and James Adam, *1774*

Tour of the House

NORTH FRONT

Visitors enter the house through the north front. The portico was added to the original brick house by Robert Adam in the 1760s. It rises to the full height of the early 18th-century two-storey house and hides the earlier brick building. Adam's patron, Lord Mansfield, needed a larger house to accommodate his two great-nieces and to entertain in a manner consistent with his newfound status and wealth, following his appointment as Lord Chief Justice and his elevation to the peerage in 1756. The plasterwork, including the 'Dorick cornice in hard stucco', was installed by Joseph Rose between 1768 and 1769.

The 2nd Earl of Mansfield instructed his architect, George Saunders, to add the two wings, which were made in white Suffolk brick and were never toned to match the main body of the house. Today, the north front has been restored to look as it did during the time of the 2nd Earl in the 1790s, with a sand and paint finish delicately scored into rectangles to look like stone.

SOUTH FRONT

Adam was clearly proud of the grand south front and terrace at Kenwood, as he published engravings and details of its decoration in 1774 in his book *The Works in Architecture of Robert and James Adam*. Adam added the library to the east of the house to counterbalance the orangery and explained that 'the decoration of the south front, excepting that of the west wing, is entirely new, and became in some measure necessary, to conceal the brick-work, which being built at different times, was of various colours – The Attic story is a late addition to the house.' The façade was decorated with plasterwork ornament in 1773, executed by Joseph Rose, but dismissed by a contemporary as 'no better than Models for the Twelfth-Night Decoration of a Pastry Cook'. By 1778, the south front was decorated with an oil cement known as 'Liardet's composition', of which the Adam brothers bought the patent, but which was unstable and deteriorated rapidly.

'[Lord Mansfield] gave full scope to my ideas: nor were they confined by any circumstances, but the necessity of preserving the proper exterior similitude between the new and the old parts of the buildings; and even with respect to this where the latter appeared defective in its detail I was at full liberty to make the proper deviations.'
Robert Adam, writing in *The Works in Architecture of Robert and James Adam*, 1774

Above: *View of the entrance hall, which Lord Mansfield also used as a dining room. It has been redecorated to reflect its original colour scheme*
Right: A *Country Life* *photograph of 1913, showing the entrance hall before its contents were sold in 1922. At that time, the mahogany wine cistern – designed by Adam and still in the entrance hall – was being used as a plant pot*

1 ENTRANCE HALL

The entrance hall is a modest room, reflecting the scale of the earlier 18th-century house. It originally had three doors in the south wall, leading to a parlour, a drawing room and a cupboard; the central door was blocked in 1815. It was the last in the sequence of rooms at Kenwood to be redesigned by Robert Adam between 1773 and 1774. Lord Mansfield also used it as a formal dining room on special occasions. It is likely that His Excellency Thomas Hutchinson (see page 41), exiled loyalist Governor of Massachusetts, dined in this room in August 1779, when he commented, 'My Lord, at 74 or 5, has all the vivacity of 50'.

Adam's Designs

Adam and Lord Mansfield must have changed their minds while designing the room, as Adam's original drawing for the ceiling does not contain any references to dining (see page 7), but the final decoration and furnishing of the room clearly demonstrate that it was used as a dining room. The central ceiling panel, painted by the Venetian artist Antonio Zucchi (1726–95), shows Bacchus, god of wine, and Ceres, goddess of agriculture, characteristic subjects for dining rooms. Zucchi's bill describes how he supplied six additional two-tone ovals with figures of Bacchantes (followers of the god of wine) in the form of cameos and an oval overmantel painting of Diana resting after the hunt with her nymphs and dogs (probably removed in 1815 and now lost).

The rest of Adam's decoration and furnishings for the entrance hall continued the dining theme. George Burns's marble chimney piece (still *in situ*) was carved from a design 'of the Messrs Adam'. It is decorated in the centre 'with a Head with a Garland of Vine Leaves and Grapes with husks and Ribbons rising out'. The furniture-maker Sefferin Nelson supplied the painted sideboard table, the two painted pedestals, vases and the mahogany cistern (for keeping wine cool), according to a bill dated December 1773. They were engraved in *The Works in Architecture of Robert and James Adam* and can also be seen in a sketch by Adam (see page 40). Some of these items have been brought back to Kenwood following their sale in 1922 but the vases as well as the '2 oval pier glasses, in white border'd frames with carved ornaments', described in the inventory of 1796, have not been returned.

The original bill of painter–decorator George Steuart (c.1730–1806), dated 1773–4, describes how the ceiling was painted 'two Greens purple & ornaments white', a colour scheme sympathetic to Bacchus's vine leaves. The room has been repainted to reflect the original decorative scheme.

2 GREAT STAIRS

The Great Stairs, built between 1767 and 1769, were originally lit by a window on the first-floor landing and decorated with a ceiling designed by Robert Adam in 1767. Following the construction of the dining room in the east wing in 1796, the landing window was removed and a skylight inserted. The main door opening off the staircase still leads into the antechamber, but two other doors, originally leading to a small parlour in the north wall and a back court and water closet in the east wall, no longer serve these spaces.

As elsewhere at Kenwood, Joseph Rose supplied the plasterwork, and John Minshull the carved work. Sefferin Nelson carved at least one pedestal to carry a glass lantern, following a design by Robert Adam. The drawing was annotated, 'A Term as wanted for the Great Staircase at Kenwood. Mr Nelson is to make one Compleat, & if that is liked he is to do 3 more'.

The balusters, 'maid according to the Hony Suckel [anthemion] pattern', were supplied by William Yates in 1769. Paint research has shown that they were originally painted a mid-blue colour, which has been restored.

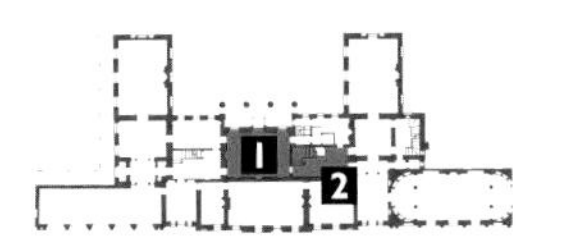

Below left: *View of the Great Stairs. The balustrade has been repainted a mid-blue colour, as it was in Adam's original decorative scheme*

Below: *Design by Robert Adam for a pedestal made by Sefferin Nelson for the Great Stairs at Kenwood, c.1768–9*

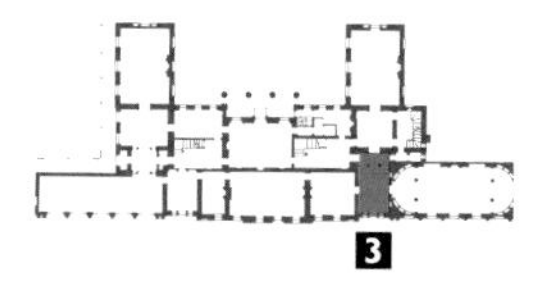

Right: A Country Life photograph of the antechamber in 1913. The wall niches were being used to display ceramics at this time
Below: View of the antechamber today, with the statues reinstated in the niches and the original long stools, which have been returned to Kenwood

3 ANTECHAMBER

Together with the library, the antechamber was the only room to be added to the ground floor of the house by the Adam brothers. It served as a vestibule to the library and originally contained three niches for statues, a screen of columns and a Venetian window, giving a view out over London. The proportions of the antechamber carefully counterbalanced those of the Housekeeper's Room to the west, providing Kenwood with its impressive, symmetrical south façade.

According to the inventory of 1796, the 'Vestibule' was furnished with '2 white Plaister whole length figures, in Niches' and included four carved, white-painted long stools, with loose seats upholstered in blue leather. These were supplied by a leading London carver, possibly Nelson, and were missing until recently. Two of these have been acquired with the help of the Art Fund and are now on display here. The seating also included six black japanned chairs, with caned seats and six mahogany Chinese elbow chairs.

Sculptures

An early scheme for the north wall of the antechamber, proposed by James Adam in about 1764, shows one of the statues set in a niche. The sculptures were plaster casts supplied by James Hoskins and Samuel Oliver in 1771, and represented the Roman statue of Flora; 'Teis' (probably the goddess Thetis, a sea nymph); and a muse. The plaster copies recently installed at Kenwood are casts of Flora and a muse, probably Euterpe, muse of music, taken from marble statues in the dining room at Syon House, Middlesex. The originals at Syon were sourced in Rome by James Adam in 1761 for his patron, the Duke of Northumberland.

Decoration

The antechamber was decorated by the carver John Minshull and the plasterworker Joseph Rose, who supplied a decorated plaster ceiling moulding to a design by Robert Adam, described as an 'Ornam[en]ted Fann with Festoons &c in Center of Ceiling 9 feet diameter'. The antechamber has been redecorated to reflect its original colour scheme, with pale green lead oil paint on the walls and white columns, which had been painted red porphyry in 1815.

The Adam Company

Although Robert Adam (1728–92) was the most famous member of the Adam family, his father, William (1689–1748), founded the original architectural practice in Scotland.

It was an unusually large family practice, for which Robert and his three brothers originally worked. Robert's oldest brother, John, took over the running of the firm on William's death in 1748, with Robert as a partner. In 1754, however, Robert left Scotland to go on a Grand Tour and on his return in 1758 he settled in London, where he set up his own architectural practice. His Grand Tour experience, particularly his visit to the palace of the Roman Emperor Diocletian in Split in Croatia, led him to introduce neoclassical motifs and forms into his work, which gained great popularity in late 18th-century England.

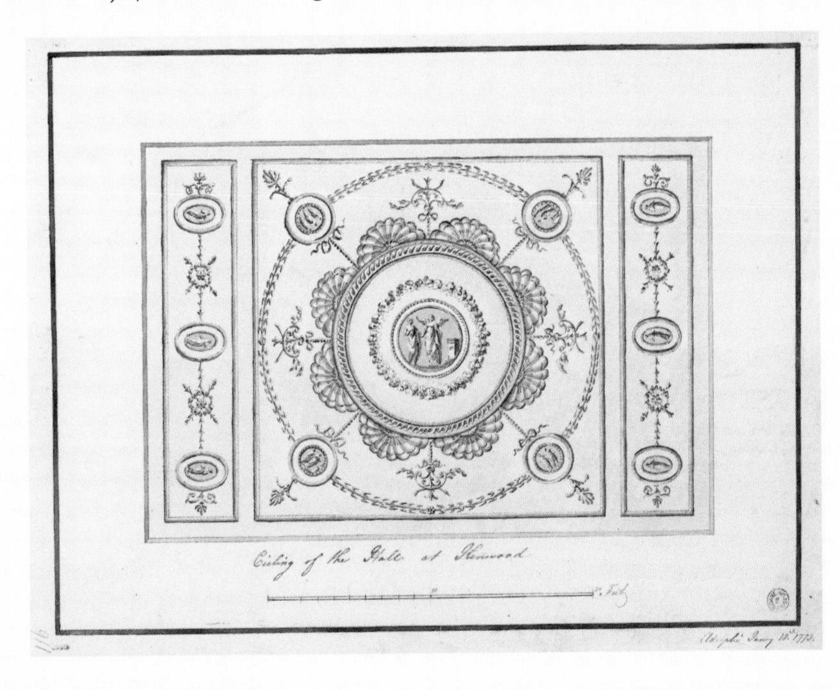

Left: *A design for the entrance hall ceiling by Robert Adam, 1773. The modified final design was executed by the plasterer Joseph Rose*

Below left: *Portrait of Robert Adam, c.1770–75, attributed to George Willison. Robert Adam was the lead architect in the family firm*

Below: *Portrait of James Adam by Allan Ramsay, 1754. James Adam created early designs for Kenwood and is responsible for the antechamber*

Robert was joined in London by two of his sisters and his brothers James (1732–94) and William (b.1738). John, the oldest brother, remained in Scotland, where he ran the architectural practice inherited from their father. Robert became the principal director of 'William Adam & Co', a firm of developers and builders' merchants set up in 1764. The company's ambitious projects included the building of the Adelphi on the Embankment behind the Strand in London and the management of brickworks in London and Essex. Despite many financial difficulties and the strains these caused between Robert and John, the firm continued operating until 1801. Robert and James Adam kept separate accounts for the profits of their joint architectural practice.

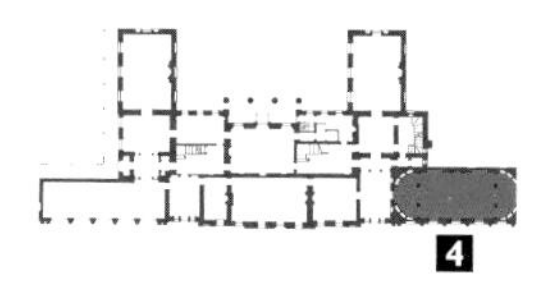

4 LIBRARY

The new library or 'Great Room' at Kenwood was built between 1767 and 1770 by Robert Adam and his craftsmen. It is one of the most famous of Adam's neoclassical interiors and represents the grand climax of the guest route through the house. Adam proudly illustrated it in 1774 in *The Works in Architecture of Robert and James Adam* and explained how 'the great room, with its anti-room was begun by Lord Mansfield's orders in the year 1767 and was intended both for a library and a room for receiving company'. The unusual shape of the room – a double cube with semicircular apses and a coved ceiling – was inspired by Adam's interest in antiquity and by his travels in Italy, the ceiling in particular 'in the form and style of those of the ancients'.

Twelve designs for the library, signed by Robert and James Adam, dating from between 1764 and 1767, survive in Sir John Soane's Museum. Three of these show the wall elevations and include designs for furniture. Adam usually only supplied furniture designs if the client specifically requested them, but his drawing of the north elevation for the 'Great Room' shows two alternatives for sofas and frames in the mirrored recesses (see page 38). Samuel Curwen, a visitor at the time, observed the 'very beautiful glasses in gilded frames, one being the largest plate I have ever seen, of French manufacture, not less in my conjecture than seven feet in height and 3½ in breadth'.

The drawing of the south elevation (see page 10) shows Adam's designs for three carved and gilt window cornices with festoon curtains and for the pair of pier glasses, made by William France between 1769 and 1770 (the originals are still *in situ*). William France, a furniture-maker and royal upholsterer, provided the furniture, some of which followed Adam's drawings and included a pair of eight-legged pier tables and 'three scrole headed sopha frames for the windows [window seats]', for which he was paid in March 1770. Adam also probably supplied the design for the marble chimney piece, carved by John Devall, and completed in 1769. Three window seats have been reinstated in the window bays to evoke the arrangement of the south wall, as designed by Adam; the one in the centre dates from the 18th century.

Right: *View of the library, which has been redecorated in Adam's original colour scheme. Adam designed the window cornices and the pier glasses*

Below: *A portrait of Lady Mary Coke by Edward Harding, after Allan Ramsay, 1798*

'I dined today at Kenwood. The improvements since I saw it are very great: Ld Mansfield has laid out a vast deal of money & with a very good taste. The great room [library] he has built is as fine as it can possibly be; no expense spared. 'Twas late before I came home.'
Lady Mary Coke,
22 July 1770

Right: *A design for the south wall of the library by Robert Adam, c.1767, showing a combination of decorative plasterwork, paintings, pier glasses, books and window furnishings*
Below: *Adam's designs for the two apses in the library, c.1767, which were used to display Lord Mansfield's book collection*

The ceiling is decorated with a series of 19 oil paintings on paper by Antonio Zucchi. The scheme was probably devised collectively by the artist, architect and patron. The central image shows the demi-god Hercules between Glory and the Passions – a subject alluding to the wise judgement of the patron, Lord Mansfield. The four roundels represent the four seasons; the four square panels contain symbolic figures of Theology, Jurisprudence, Mathematics and Philosophy; the vignettes (half-circles) depict Justice embracing Peace, Commerce, Navigation and Agriculture; and the apses are decorated with five classical subjects. Over the door is a painting of the infant Hercules strangling two serpents. The only other decorative painting hung in the library was a full-length portrait of Lord Mansfield over the chimney piece. Painted by the Scottish artist David Martin in 1776, the year in which the sitter became an earl, it replaced the classical judgement scene proposed in Adam's 1767 drawing. The original portrait was moved in 1922 to Scone Palace, Perth, ancestral home of the Mansfield family (the painting here today is a modern copy).

Work in the 'Great Room' must have been well advanced by December 1769, when the plasterer Joseph Rose submitted his bill for 'the inside of Library'. It was probably completed by July 1770, when George Steuart was paid 'for Painting done at Library & Anti R Kane Wood from Jan[uar]y 1 1768 to June 1770'.

Left: *Photograph of the library, taken for* Country Life *in 1913, showing much of the original furniture still in place. The sofa was designed by Robert Adam and made by William France in 1769. The mirrored niches by the fireplace were turned into bookshelves by the 3rd Earl of Mansfield in the early 19th century*

Below: *The central ceiling painting in the library by Antonio Zucchi, 1769, depicts the story of Hercules resisting temptation, with the sculpture of 'Glory' in the high temple beyond, alluding to Lord Mansfield's own good judgement*

The 'Great Room' was the largest room at Kenwood and it was immediately admired by contemporaries. *The Ambulator; or the Stranger's Companion in a Tour round London* (1782) described how 'the new room latterly built by his Lordship, from a design of Mr Adams, is considered, by architectural judges, as well for its proportions and decorations as its novelty to be superior to any thing of the kind in England'.

Lord Mansfield was a great collector of books and had been a good friend of the poet Alexander Pope. According to an inventory of books moved from Kenwood between 1831 and 1840, the library contained a 1504 edition of Petronius's *Satyricon* and Jean-Jacques Barthélemy's *Natural Philosphy* of 1795. The room was always intended 'for receiving company'. In common with other late 18th-century libraries, it was used by the family to host dinners, perform music, play games and pass the evening.

Recent research has shed new light on the original paint colours and decorative scheme in the library. Paint analysis was carried out by taking cross sections through the paint layers. Tests showed that there was originally very little gilding in the library. At frieze level, for instance, the decoration of lions and deer heads was highlighted with lead white oil paint and there was no gilding on the ceiling, which was painted with a varied palette of greens, blues and pinks. This decorative approach was typical of Robert Adam's painted colour schemes in the period between 1765 and 1770. Unfortunately, this original scheme did not last long, as, probably some time between 1795 and 1815, a new gilded scheme was introduced, possibly as part of the 2nd or 3rd Earl of Mansfield's alterations to the house.

Right: A Coast Scene with Fishermen Hauling a Boat Ashore *by* J M W Turner, *c.1803–4*
Below: *View into the dining room lobby, with the dining room beyond*
Below right: View of Dordrecht *by* Aelbert Cuyp, *c.1655*

5 DINING ROOM LOBBY

Described in the inventory of 1796 as the 'Anti Room adjoining [the] Dining Room', this small lobby was part of the new east wing commissioned by the 2nd Earl of Mansfield between 1793 and 1796. The 2nd Earl, who had been ambassador to Vienna and Paris, instructed the architect George Saunders (c.1762–1839) to make significant alterations to the house, including the addition of a dining room. This lobby fulfilled an important practical function: it was a place where the servants could assemble dishes before carrying them into the dining room next door, and it had access to the stone service staircase, service wing and kitchens through the two doors in the east wall. Occupying the site of the 'Back Court' on Adam's floor plan of 1767, it was originally painted blue and white, but in 1815 it was decorated with graining to resemble wood. The inventory of 1796 shows that the lobby was furnished with a mahogany sideboard 'in a square carv'd white painted frame, and brass Guard rail, Honeysuckle fret on top', beside two carved and painted pedestals and vases (previously in the entrance hall), standing on an eight-foot long piece of oil cloth. Two mahogany dumb waiters were also on hand to hold dishes, while a full-length plaster figure (originally from the niche in the north wall of the antechamber) decorated the space.

This wing was redecorated in 2000 and a floorcloth was fitted – a precursor of linoleum, which is made up of layers of oil paint on canvas. The two carved and gilt mirrors were brought back to Kenwood in 1994; they are original to Kenwood and pre-date Robert Adam's work for Lord Mansfield. The Cuyp painting below was one of Lord Iveagh's first purchases from Agnew's.

Lord Iveagh as a Collector

Edward Cecil Guinness, 1st Earl of Iveagh (1847–1927), bought Kenwood in 1925, planning to give it to the nation with part of his collection of paintings.

When Guinness Breweries became a public company in 1886 and Edward Cecil Guinness was no longer sole proprietor, he began to build up his picture collection. He had a number of houses in Ireland and England and in 1887 he and his wife, Adelaide, acquired a London town house at 4–5 Grosvenor Place, which needed furnishing. Between 1887 and 1891, he bought over 200 paintings, largely from the Bond Street dealers Thomas Agnew & Sons. On his death in 1927, Lord Iveagh bequeathed 63 paintings to Kenwood, which reflect the quality of his collection. Many of the works he acquired came from established collections, with distinguished provenances, such as Claude de Jongh's *Old London Bridge* (1630), once in the Marquess of Exeter's collection at Burghley House. He also bought portraits of children and aristocratic or famous women by important 18th-century artists, notably Sir Joshua Reynolds, Thomas Gainsborough and George Romney. Among the most popular works on display are Reynolds's *Mrs Musters as 'Hebe'* (1782); Gainsborough's *Mary, Countess Howe* (c.1764); and Sir Thomas Lawrence's *Miss Murray* (1824–6). As a sportsman and sailor, he collected hunting and marine scenes, including 'The Iveagh Sea-Piece', Turner's celebrated *A Coast Scene with Fishermen Hauling a Boat Ashore* (c.1803–4). His most expensive purchase was the Rembrandt self-portrait (see title page), acquired in 1888 for £27,500 together with the *Portrait of an Unknown Woman* by Ferdinand Bol, once believed to be by Rembrandt.

Through Lord Iveagh's generosity, visitors can now enjoy a gallery of internationally important art, free of charge.

Above: *Detail of* Old London Bridge *by Claude de Jongh, 1630. This painting once belonged to the Marquess of Exeter at Burghley House*

Left: *Lord Iveagh collected many female portraits from the 18th century, including this portrait by Sir Joshua Reynolds of* Mrs Musters as 'Hebe', *the Greek goddess of youth, 1782*

Below: *The 1st Earl of Iveagh, who bought many paintings from Agnew's between 1887 and 1891, photographed by Walter Stoneman, 1926*

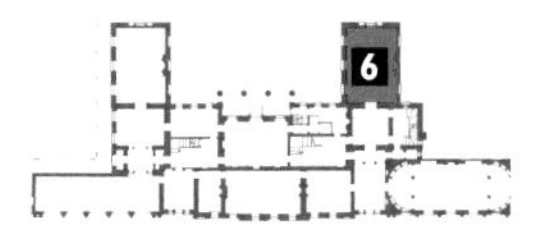

Right: *View of the dining room, which contains some of the most celebrated paintings in the Iveagh Bequest*
Below right: The Guitar Player *by Johannes Vermeer, c.1672, which Lord Iveagh bought for £1,050 in 1889*

6 DINING ROOM

The 2nd Earl of Mansfield wanted to have a dedicated dining room at Kenwood, rather than using the entrance hall, where the 1st Earl of Mansfield had entertained his guests with formal dinners. Plans for the dining-room wing may already have been in place before 1793, when the architect Robert Nasmith was commissioned by the 2nd Earl, but Nasmith's death in August that year meant that another architect, George Saunders, had to carry out the work.

According to the inventory taken on the death of the 2nd Earl in 1796, the dining-room furnishings included two mahogany sideboards, next to Chinese vases on top of pedestals, one of which served as a plate warmer, and an oval wine cooler. Dinner guests sat round a large mahogany dining table, which could extend to 18 feet in length. Overlooking the scene were two marble busts of William Murray, 1st Earl of Mansfield, mounted on painted circular pedestals. The marble chimney piece, carved by John Bingley in about 1795, is decorated with the head of Bacchus and vine leaves. This room was redecorated in 2000 to evoke an early 19th-century interior, with the wall colour acting as a suitable backdrop for the Old Master paintings in the Iveagh Bequest.

Old Master Paintings in the Dining Room

The dining room contains internationally important pictures by Dutch and Flemish Old Master painters.

One of the most famous works at Kenwood is Rembrandt's *Portrait of the Artist* (c.1665), perhaps his best late self-portrait, painted when he was about 60 (see title page). Rembrandt depicts himself at work, wearing a white linen cap, holding his brushes, palette and mahlstick, with his easel at the right of the canvas. The shapes in the background possibly relate to an attempt to paint the perfect circle. The artist's serious gaze suggests his concentration and may reflect his personal troubles, as he was struggling with bankruptcy.

Nearby hangs a portrait by Frans Hals of Pieter van den Broecke (1633) as well as portraits by Anthony van Dyck of James Stuart, 1st Duke of Richmond and 4th Duke of Lennox (c.1636), and of Princess Henrietta of Lorraine with a black page (1634). The fluid brushstrokes of Hals's portrait of his friend, a merchant seaman from Antwerp, who was celebrating a successful career with the Dutch East India Company, contrast with the smooth elegance of Van Dyck's portraits. Van Dyck's sitters are both aristocrats; Stuart was cousin to King Charles I and Henrietta was widow of Louis de Guise, Prince of Pfalzburgh. Lord Iveagh bought Johannes Vermeer's *The Guitar Player* (c.1672) in 1889 for £1,050, when the artist's popularity was increasing. Relatively few works by Vermeer survive. His subject matter often contains musical scenes and he is admired for his handling of light.

Above: Pieter van den Broecke *by Frans Hals, 1633. This informal portrait of the artist's friend and merchant seaman Pieter van den Broecke was bought by Lord Iveagh when Frans Hals was considered second only to Rembrandt as a Dutch portrait painter*

Left: *Portrait of James Stuart, 1st Duke of Richmond and 4th Duke of Lennox, by Sir Anthony van Dyck, c.1636, bought by Lord Iveagh when Van Dyck was popular*

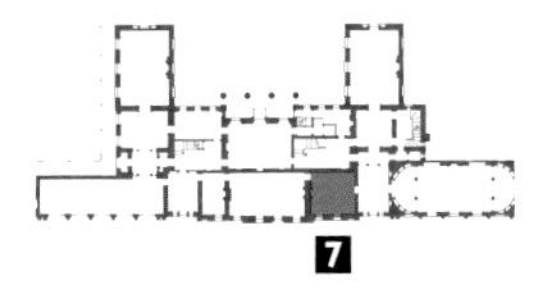

Below right: Design for the chimney piece in Lord Mansfield's Dressing Room by Robert Adam, c.1779–80. The marble fireplace survives and closely follows the original design

SOUTH FRONT ROOMS

The south-facing rooms along the terrace were originally part of the earlier house, dating from the early 18th century. They were later remodelled by Robert Adam and consisted of four family rooms: a drawing room, a parlour for informal dining and dressing rooms for both Lord and Lady Mansfield, which were situated directly below their first-floor bedrooms. Adam intended to show the alterations to these rooms in a later edition of his *Works in Architecture of Robert and James Adam* but this was never published. Accounts show that Adam's additions included chimney pieces, cornices, dado rails and shutters.

The four rooms are flanked by the antechamber and Housekeeper's Room, with the library and orangery beyond. This long, symmetrical south façade is both elegant and impressive, looking over the grounds towards central London.

The rooms were subsequently altered by the 3rd Earl's architect, William Atkinson, who combined the central rooms into the 'Book Room' in 1815, later known as the 'Breakfast Room'. The room was lined with books and was used as a study. These rooms were modified before the house was opened to the public in 1928; some of the doors were moved to allow easier access for visitors.

The south front rooms are displayed as domestic interiors of the late 18th century and are hung mainly with Georgian paintings from the Iveagh Bequest. The curtains and carpets have been informed by the inventory of 1796 and reflect the appearance of an 18th-century interior.

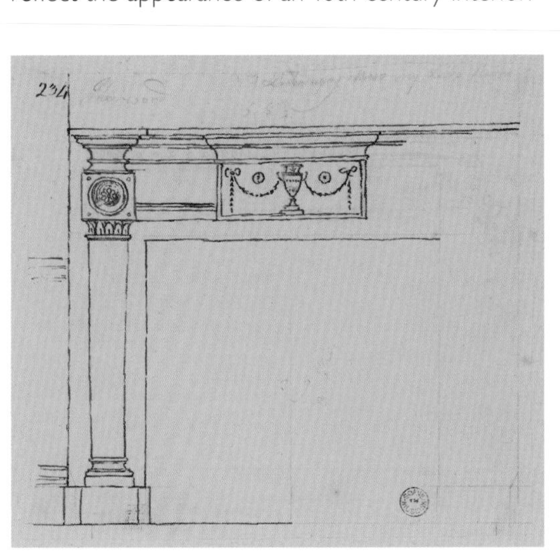

7 LORD MANSFIELD'S DRESSING ROOM

This room served as a study for the 1st Earl of Mansfield and is known to have contained a marble bust of Homer (now at Scone Palace), given to him by his mentor, the poet Alexander Pope. As would be expected in a gentleman's dressing room, the room held some of the earl's most valued possessions and was hung with portraits and drawings.

In the mornings, Lord Mansfield would have come downstairs from his bedroom above and been dressed by his manservant in this room. The library opposite was too large to heat for daily

***Left:** Lord Mansfield's Dressing Room, which is hung with paintings from the Iveagh Bequest*
***Below:** Marble bust of Homer by Joseph Wilton, c.1760, on display in the library. Lord Mansfield was given a similar bust of Homer by his mentor, the poet Alexander Pope, once displayed in this room, and now at Scone Palace, Perth*

use, so this more modestly sized room served for reading and receiving visitors. This room links the reception suite with the family's domestic apartments.

Three Adam drawings relating to the marble chimney piece survive in Sir John Soane's Museum, London. The deep exterior wall of the pre-Adam house can be seen by the mahogany door from the antechamber. In the 1st Earl's day there was a doorway to the left of the chimney piece and one on the north wall to the Great Stairs. The current position of the doorway to the Breakfast Room dates from the 1920s, when Kenwood was adapted for public use.

This room was subsequently used by Louisa, 2nd Countess of Mansfield, and listed in 1796 as the 'Yellow Room' with its '2 yellow vine leaf festoon window curtains, lin'd & fring'd'. It was renamed the 'Japan Room' in 1817, when it was redecorated by William Atkinson for Frederica, the 3rd Countess. By 1831, it contained black and gold lacquered furniture in the chinoiserie taste and 26 paintings, including David Wilkie's famous painting *The Village Politicians*, now at Scone Palace. Today it contains paintings from the Iveagh Bequest, including a portrait of the 1st Earl of Iveagh by H M Paget (1856–1936), after A S Cope.

Above: *The Breakfast Room, which now contains paintings from the Iveagh Bequest*
Right: *A photograph of the Breakfast Room, taken in 1913 for* Country Life, *showing the room in use by the Grand Duke Michael. The bookshelves were installed in 1815 by the 3rd Earl of Mansfield, who used this room as a library*

8 BREAKFAST ROOM

The Breakfast Room was originally two rooms, the drawing room and parlour, both directly accessible through doors from the entrance hall. The parlour served as a family dining room and the drawing room as a room to withdraw to after meals. The shutters and the chimney piece date from the 18th century and are surviving remnants of Robert Adam's scheme.

By 1796 the rooms were known as the 'Outer and Inner Library' with both rooms having the same 'blue silk Damask pulley rod window curtains, scollopt & valens', which have influenced the current curtain hangings. The outer library contained an oval pier glass with 'an ornamental carv'd & gilt frame' made by the craftsman George Burns as part of earlier furnishings designed by Adam; the design was illustrated in his published book, *The Works in Architecture of Robert and James*

Adam of 1774. The rooms were combined in 1815 by the 3rd Earl's architect, Atkinson, as a 'Book Room' and bookcases were fitted to the walls. At this time, many changes were made to the room, with extensive repairs primarily due to dry rot. It was called the 'Breakfast Room' by 1831, with large mirrors on the pier walls, with three pier tables 'inlaid and ornamented with brass on 8 legs'. Today, the room is hung with paintings from the Iveagh Bequest, notably a view of Hampstead Heath by John Constable and Gainsborough's *Two Shepherd Boys with Dogs Fighting*.

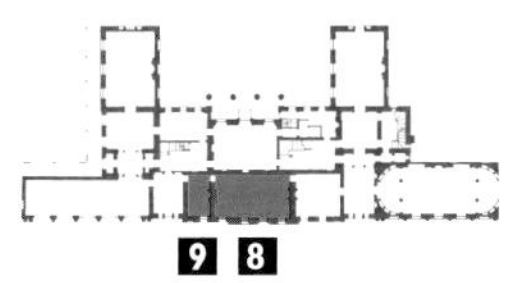

Left: Hampstead Heath with Pond and Bathers *by John Constable, 1821*
Below: *Lady Mansfield's Dressing Room, which now contains paintings and furniture from the Iveagh Bequest*

9 LADY MANSFIELD'S DRESSING ROOM

The 1st Countess of Mansfield used this room for dressing, receiving company and reading. The daughter of the 2nd Earl of Nottingham, Lady Betty was 34 years old when she married the young lawyer, William Murray, whom some deemed below her station. Lady Mary Wortley Montagu commented about their marriage, 'people are divided in their opinions, as they commonly are, on the prudence of her choice'.

This must have been an impressive room, as the antiquarian and politician Horace Walpole noted in a letter to the Reverend William Mason in 1776, 'a Theban harp, as beautifully and gracefully designed as if Mr Adam had drawn it from Lady Mansfield's dressing room, with a sphinx, masks, a patera, and a running foliage of leaves'. In 1770 the joiner Edward Lonsdale was paid for fitting bookcase doors. There was a chimneyboard 'painted by Zoochi' (Antonio Zucchi) noted in this 'Small Study' in the inventory of 1796, as well as 'An India Chintz festoon window curtain', which has influenced the modern window dressing.

The two doorways are a later addition, in an area which was originally a small, separate chamber or closet. In Adam's unpublished plan this was called the 'China Closet', presumably used for displaying Lady Mansfield's porcelain collection. The present wooden mantelpiece is a later copy of the one in Lord Mansfield's Dressing Room. By the time of the renovations of 1817 this room had become the 3rd Earl's study. The current display of paintings from the Iveagh Bequest includes *Old London Bridge* by De Jongh and a portrait of Louis, Duc de Bourgogne, by Rigaud and Parrocet.

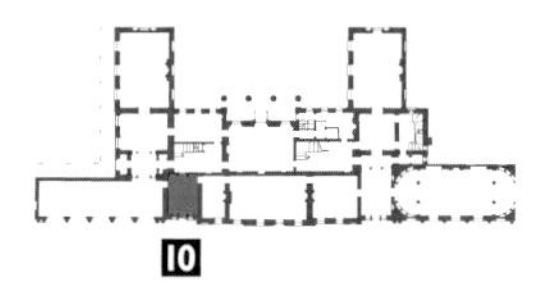

Right: *The Venetian window in the Housekeeper's Room mirrors the window in the antechamber on the south façade, designed by Adam*
Below: The Embroiderer *by Jean-Baptiste Siméon Chardin, c.1773, showing a servant absorbed in her work. Lord Mansfield's housekeeper was accused of stealing fruit in 1773*

10 HOUSEKEEPER'S ROOM

The housekeeper was a senior servant in the house and this room would originally have been close to the kitchen and service rooms, before the wings were added to the house in the 1790s. At this time the orangery was not linked to the house and it was only accessible from the garden.

By 1796, the inventory lists this space as a 'Reading Room' and its contents include eight maps, two telescopes and a small globe. In 1817 its use had changed to a 'school room', at a time when the 3rd Earl of Mansfield's numerous children would have been growing up at Kenwood. A section of an earlier and unusual hand-painted decorative wall scheme from this period has been partially uncovered above the central window. The Venetian or 'Palladian' window provides symmetry on the external façade, balancing with the antechamber.

Theft at Kenwood

In 1773, the *Morning Chronicle* newspaper reported that Lord Mansfield had noticed that fruit was going missing from an apple tree. His gardener was instructed to lie in wait one evening and he caught the housekeeper red-handed. 'Having filled her apron, she was about to de-camp, when the gardener ran with fire and fury at her, and drawing his knife, cut down her apron… down tumbled the apples, pears and plumbs [sic] from the lap of plenty.' In the interview, Lord Mansfield noted that his housekeeper 'has served me for a long time very faithfully' and she was not discharged.

Children at Kenwood

Generations of children have lived, worked and played at Kenwood, including the Mansfield family, estate staff and house servants.

John Stuart, 3rd Earl of Bute (1713–92), lived at Kenwood between 1746 and 1754; his ninth child, Charles, was born here in January 1753. The 1st Earl of Mansfield and his wife did not have children but brought up their great-nieces, Dido and Elizabeth. The arrival of the girls at Kenwood, aged five or so, would account for the addition of a third floor to the house for nursery accommodation in about 1766.

An account book of pocket expenses, dating from 1785 to 1790, belonging to Lord Stormont (who was later the 2nd Earl of Mansfield), lists various amounts for children's shoes and toys. He paid for a dancing master for William in 1782 and gave him money 'on his return to school' in August 1790 (£.1 1s. 0d.).

Some of the 2nd Earl's six children are listed in the household inventory of 1796, with a portrait of his daughter Lady Caroline specified, as well as the contents of her bedroom and sitting room in the west wing of the house. Her teenage brothers, the Hon. Charles Murray and the Hon. George Murray, had bedrooms in the service wing at this time.

Kenwood would have been a lively place with the 3rd Earl's family of nine children who lived here from 1796 to 1840. Two of his sons, David and Charles, aged seven and eight, were tutored by a Thomas Roy in 1818, at a time when there was a 'School Room' located on the ground floor.

The dairymaid's son, John Elliot, is known to have grown up on the estate in the 1830s; Frederica, 3rd Countess of Mansfield, painted John's portrait when he was a child. Within the household accounts, children also appear on the staff. The housemaid, Martha, is listed in the accounts of 1786 and she was probably 13 or 14 years old. She was paid £8 a year. A steward's room boy, George Pike, is also listed in 1835.

Peruke Makers D°. 3 3 2
To Dido on her Birthday by L M's order . . . 5 5
Machine for measuring Rainwater 1 11 6
Feeding & Fatting Fowl . . . . . P. . . 68 17 10
Letters & Turnpikes 1

Above: *Portrait of Lady Caroline Murray by an unknown artist, c.1790. She was the daughter of the 2nd Earl of Mansfield and is known to have had rooms in the west wing at Kenwood*

Above left: *Detail from Anne Murray's account book, covering the period from 1785 to 1793, which shows that Dido Belle was given five guineas on her birthday*

Left: *Portrait of the three sons of John, 3rd Earl of Bute, by Johann Zoffany, c.1763–4. Charles, who is balancing on the tree, was born at Kenwood in 1753*

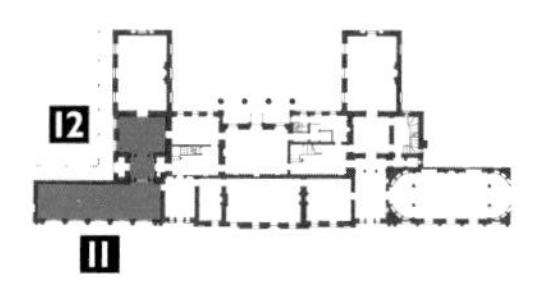

Right: *The large south-facing windows in the orangery give splendid views over the estate and help warm this space, which once contained exotic plants*

Below right: *Engraving of a 17th-century Dutch orangery by Jan Commelin from* De Nederlandze Hesperides. *The stove is clearly visible on the front right and the room contains a variety of plants*

Facing page left: The Cherry Gatherers *by François Boucher, 1768, on display in the Green Room*

Facing page right: *View of the Green Room today, looking towards the orangery. The frieze incorporates gilded lyres, dedicating this wing to Apollo, god of music and poetry*

Facing page bottom left: *Watercolour of the Green Room, probably painted by one of the 3rd Earl of Mansfield's daughters, in 1824*

Facing page bottom right: *French boulle clock, c.1700, probably bought by the 3rd Earl of Mansfield*

11 ORANGERY

The orangery was originally a separate, freestanding building, accessible only from the garden. Although it was not shown on Rocque's map of 1745 or on Mary Delany's Kenwood sketch of 1755 (see page 35), it was probably added between 1746 and 1754 by the owner, John Stuart, 3rd Earl of Bute, a keen horticulturalist. The surviving cornice dates from this period.

This room was called the 'Green House' on Adam's plan of 1774, with 'Orangerie' in the French subtitle. The exterior was refaced by Adam to make it symmetrical with the library at the other end of the south front, but the earlier brickwork survives underneath. It would originally have had a flagstone floor.

As well as housing plants and providing impressive views towards London, the orangery would have been used by the family for indoor walks on rainy days. Orange and peach trees, as well as other exotic plants, would have survived due to the warmth from the nearby kitchen and bakehouse. In an article about Kenwood in 1780, *The Ambulator* stated, 'the green house also is superb, and contains a very large collection of curious and exotic plants, trees & c'.

The orangery was linked to the house in the 1790s with a set of large glazed doors when the Music-Room wing was added. At this time, the plants were warmed by stoves, and by an underfloor heating system of hot-water pipes, still visible by the window. The window frames and glass were replaced between 1815 and 1817.

12 GREEN ROOM

When David Murray, 2nd Earl of Mansfield, inherited this house from his elderly uncle in 1793, he added this wing to increase the family accommodation. The kitchen and service quarters were previously located here but they were moved to a more discreet position to the east of the house. This wing links to the main house and the orangery and provides access to the flower garden. The room was simply referred to as 'Anti Room off Music Room' in the inventory of 1796.

Its current name first appeared in 1831 and appears to have come from the wall and column colour. The first scheme by George Saunders was a light green colour, with off-white joinery. A darker, mid-green wall colour was used in 1817 when it was called the 'Supper Room'. At this time the columns were painted to imitate green porphyry and the timber mouldings were painted to resemble oak.

The current scheme was informed by a watercolour view of 1824, probably painted by one of the 3rd Earl of Mansfield's daughters. The columns frame the view of the grounds seen through the orangery windows, while the west-facing windows look out to the flower garden and the landscape beyond.

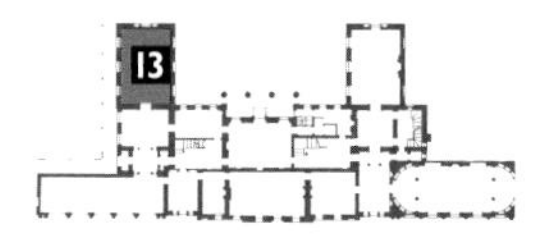

Right: *View of the Music Room, which was added by the 2nd Earl of Mansfield between 1794 and 1796*
Below middle right: *A photograph of the Music Room, taken for Country Life in 1913*
Below bottom right: *This scene of putti picking hops, 1794–7, by Julius Caesar Ibbetson, is a surviving fragment of the original Music Room decoration, removed from the walls in 1926*
Below: *Portrait of Joseph Farington, 1795, by Sir Thomas Lawrence*

13 MUSIC ROOM

This room was called the 'Music Room' in the inventory of 1796, shortly after its completion. It now contains some of the most famous paintings from the Iveagh Bequest, including Gainsborough's *Mary, Countess Howe* and works by Reynolds, Romney and Hoppner. Ladies would withdraw here after dinner, joined later by the gentlemen, for tea, music and a view of the gardens. The dairy was also built at this time, between 1794 and 1795, on the brow of the hill past the flower garden, and it would have been visible from the west-facing windows, which catch the afternoon sunlight.

The artist Julius Caesar Ibbetson (1759–1817) worked at Kenwood between 1794 and 1797 and provided a decorative scheme for this room, probably commissioned by Louisa Cathcart, 2nd Countess of Mansfield, which included painted scenes of Welsh castles and views, but predominantly scenes of *putti* performing various agricultural tasks. Ibbetson's scheme was removed in 1926 to allow enough space for Lord Iveagh's painting collection but sections of his work survive above the organ and over the door.

The organ by Robert and William Gray was sold in the 1920s and has been replaced with an instrument by John England and Son of a similar date and in an identical position, which is played for recitals. By 1831, much of the fine furniture listed in this room was French in origin. The Green Room and the Music Room were redecorated in 2000 with a scheme influenced by historical research, to complement the painting and furniture collection.

'[Ibbetson] gave his employers a great deal of trouble – there was no depending on him.' The artist Joseph Farington, writing in his diary in 1799

Masterpieces in the Music Room

The Music Room contains some of the greatest British portraits of the 18th century, including works by Gainsborough, Reynolds and Romney.

Lord Iveagh was particularly fond of collecting female portraits of the 18th century and his accounts with the London art dealer, Agnew's, between 1887 and 1908 reveal that he regularly spent large sums of money on this type of painting.

Thomas Gainsborough

Mary, Countess Howe is the best-known painting of this period in the collection. Gainsborough was one of England's leading portrait painters of the 18th century and was renowned for his full-length portraits of society ladies. This portrait was painted in about 1764, when Richard Howe and his wife, Mary Hartopp, were taking the waters in the spa town of Bath and commissioned their portraits, probably for their London home.

Sir Joshua Reynolds

There are 17 works by Sir Joshua Reynolds (1723–92) at Kenwood, *Mrs Musters as 'Hebe'* depicts the sitter as the Greek goddess of youth. *Kitty Fisher as Cleopatra Dissolving the Pearl*, 1759, shows one of the best-known London courtesans of the time as the queen of Egypt. Reynolds excelled at capturing the vitality of children in portraiture and the picture of the Brummell boys playing with a puppy, painted between 1781 and 1782, demonstrates his mastery of movement and brushwork.

George Romney

Lord Iveagh also collected paintings by the fashionable English portrait painter George Romney. There are several examples in this room, including *Emma Hart at Prayer*, (c.1782–6) reminiscent of the penitent Mary Magdalen. Romney was infatuated with his sitter and muse, Emma Hart, whom he painted 28 times. Best known as the mistress of Lord Nelson, Emma was admired for her varied expressions and dramatic posturing.

Left: Kitty Fisher as Cleopatra Dissolving the Pearl *by Sir Joshua Reynolds, 1759. The* Middlesex Journal *complimented Reynolds that he had 'come as near the original as possible' in this portrait of the celebrated courtesan Kitty Fisher*

Below left: Mary, Countess Howe, *by Thomas Gainsborough, c.1764, is one of the highlights of the Iveagh Bequest*

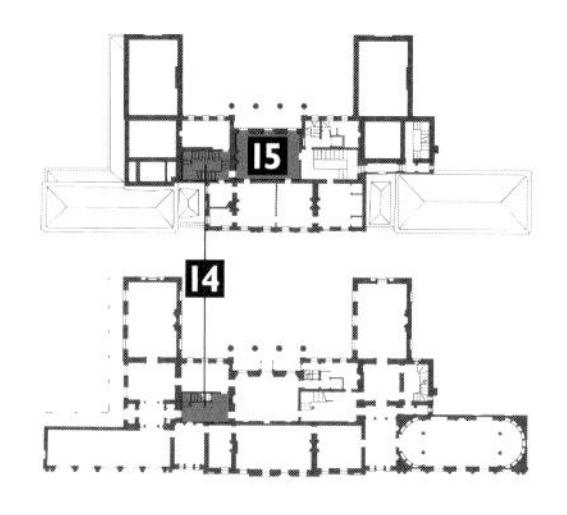

Below: *A surviving fragment of the chinoiserie wallpaper, now in store, which once hung in the upper hall, dating from about 1773*

Below right: *The upper hall, photographed fully furnished in 1913 by* Country Life, *before the contents were sold at auction in 1922*

14 DEAL STAIRS

This staircase was built between 1793 and 1796 when the two wings were added. It served as a second route for the family between floors, in addition to the main staircase by Adam and the servants' back stairs. It has been decorated to look as it did between 1815 and 1817, when the 3rd Earl of Mansfield commissioned the architect William Atkinson (1774/5–1839) to work on the house. He added a screen at ground-floor level, boxing in the space under the stairs, and introduced painted wood graining on the doors. The stair balusters were painted to resemble bronze. The floorboards are new deal boards of graduating width. The Kenwood inventory of 1841 mentions 'a yard wide Kidderminster Stair carpet abt 50 yds' with this modern version following the historic design and colours of a surviving sample from the Louth Museum. The miniature collection can be seen in the room on the mezzanine level (see page 51).

15 UPPER HALL

The upper hall was the principal reception room before Adam added the library to the ground floor. The pre-Adam doors and their brass lock mechanisms are still *in situ*.

There are no surviving first-floor plans by Robert Adam, but it would seem that this room was used as an additional reception room for Lord and Lady Mansfield's close friends, with the other rooms on this floor being used predominantly as bedrooms. In 1773 it was referred to as the 'Chinese Room' due to its remodelling by Adam in the then fashionable chinoiserie style. Elements of the chimney piece were carved by the London carver Sefferin Nelson; it contains rare painted marble tiles. A fragment of chinoiserie wallpaper, incorporating birds, butterflies and flowers, also survives.

The windows looked out across Hampstead Lane and beyond until the 1790s, when the road was diverted by the 2nd Earl of Mansfield on Humphry Repton's advice. By 1796 it was called the 'North Drawing Room' and the inventory indicates that it was well furnished, containing many paintings in gilt frames. The function of the room had changed by 1817, when it was used as a 'Billiard Room' by the 3rd Earl and his family. Today, these rooms contain works from the Suffolk Collection, a group of paintings commissioned and collected by the earls of Suffolk and Berkshire, which were gifted to the nation in 1974.

Left: *Portrait of Richard Sackville, 3rd Earl of Dorset, 1613, by William Larkin, part of the Suffolk Collection, a group of paintings commissioned and collected by the earls of Suffolk and Berkshire over a period of 400 years, which was gifted to the nation in 1974. It includes an important group of full-length Jacobean portraits by William Larkin*

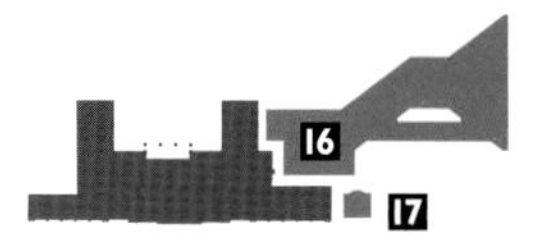

Right: *The service wing, originally called the 'office wing', was added by the 2nd Earl of Mansfield between 1794 and 1796*
Below right: *Frontispiece from* The Housekeeper's Instructor *by William Augustus Henderson, 1790s, showing staff at work in a large kitchen*

16 SERVICE WING

The service wing is made of rare purple-brown London bricks and was built to the designs of the architect George Saunders for the 2nd Earl of Mansfield and completed by 1797 for the 3rd Earl. It replaced the previous service rooms located behind the orangery, and was screened to the east of the house behind the terrace.

The service wing contains over 60 rooms, including beer and wine cellars, wash house, larder, pantry, pastry rooms and servants' bedrooms. The largest room was the kitchen, which linked directly to the dining room through a covered passage to the house. The stew stove was built in about 1796 and is one of the best surviving examples of 18th-century cooking equipment. The brick structure held a series of smouldering charcoal fires, which required great skill to control when heating the saucepans placed above. This kitchen originally had a large collection of copperware and second-best china, displayed on the three dressers. Apart from the stew stove, most of the fittings today date from the 19th century.

A variety of servants was employed to look after the Mansfield family at Kenwood. In the late 18th century, there would have been a cook, kitchen maids and scullery maids, with morning and early evening being the busiest times. The cook agreed menus with the mistress of the house, while much of the basic cooking was done by the 'upper' (or 'first') kitchen maid, who prepared meals for the servants' hall and nursery. Lower servants, such as the scullery maids, often completed tasks before the household awoke, such as lighting the kitchen range. In 1854, Mrs Freeman was the cook for the

Left: *The cast-iron open range in the old kitchen was used for spit roasting. It was made by Johnson & Ravey and dates from about 1845, when the kitchen equipment was upgraded*

Below far left: *The exterior of the bathhouse, which was fed by a spring on Hampstead Heath. Regular cold baths were believed to have health benefits*

Below left: *The stew stove, built in about 1796, is an interesting surviving example of 18th-century cooking equipment. In the inventory of 1796, 66 copper stewpans were listed*

4th Earl of Mansfield, and she was the most highly paid servant next to the butler, earning £40 a year. In comparison, the butler, Mr Barker, was paid £50; Mrs Watkins, Lady Mansfield's maid, received £25 a year; while Ann Green, the housemaid, and Margaret Morton, the kitchen maid, were both paid £12 a year.

17 BATHHOUSE

Bathhouses and private cold bathing were fashionable with the gentry from the late 17th century and were often located in a small building by the main house. This bath near the kitchens at Kenwood was originally fed from a spring, of which there were several on the Heath, accounting for Hampstead's status as a spa town in the 18th century. It was believed that regular cold baths in spring water had medical benefits and could even cure conditions such as rickets and leprosy.

The exact date of this bathhouse is unknown, but the earliest repairs are recorded in 1762. From Adam's plans of 1764 it appears that the bathhouse was rectangular with corner steps. Later improvements include apsed seats for bathers around an oval pool lined with Cararra marble, and a domed plaster ceiling. The bath was subsequently filled with silt and rubble and a limestone floor laid. When it was excavated, there was evidence of over 20 oyster shells used decoratively in the plaster walls. It was not until the 1970s, when the roof and plaster dome were rebuilt, that this building began to be restored.

The Landscape of Kenwood

Above: View of London from Highgate *by Robert Crone, c.1777. The view from the terrace at Kenwood would once have been similar to this, with visitors able to see as far as St Paul's*

THE SETTING

The woods around Kenwood House were once part of ancient woodland, possibly dating back to Saxon times. In the medieval period, they were part of the Great Park of Hornsey, the deer park of the bishops of London. The area remained woodland until the 17th century; some of the oak trees on Hampstead Heath may date from this time or even earlier.

It is clear from a map drawn by John Rocque in 1745 that even into the 18th century, the area was still undeveloped and there must have been excellent views to the city, with fashionable society coming to Hampstead to admire the view. Also visible on Rocque's map is the avenue of lime trees along the terrace, planted in about 1727. (The present trees are 20th-century replacements.)

The meandering system of paths through the grounds was laid out in the mid 18th century to provide a series of walks, ornamented with seats. Although the route of some of these walks has changed, others are probably original.

The grounds as seen today date largely from the late 18th century, when the 2nd Earl of Mansfield employed Humphry Repton to enhance the grounds. Repton imagined visitors travelling through the landscape on foot or by carriage and pausing somewhere to admire the view. He was responsible for moving the line of Hampstead Lane to the north of North Hill, to screen the house from the road and provide a sense of surprise, as first the house and then finally the landscape came into view. The 'sham bridge' on Thousand Pound Pond created a sense of distance for the viewer and evoked an Italian landscape painting, as did Wood Pond. The architect George Saunders designed the dairy, on a small hill to the west of the house, forming a focal point in the landscape.

In the 19th century the 3rd Earl of Mansfield built the trellis arbour, and the 4th Earl of Mansfield added the rhododendron garden and the veranda, allowing visitors walking round the grounds, both then and today, to enjoy the views, with the house at the centre of the landscape.

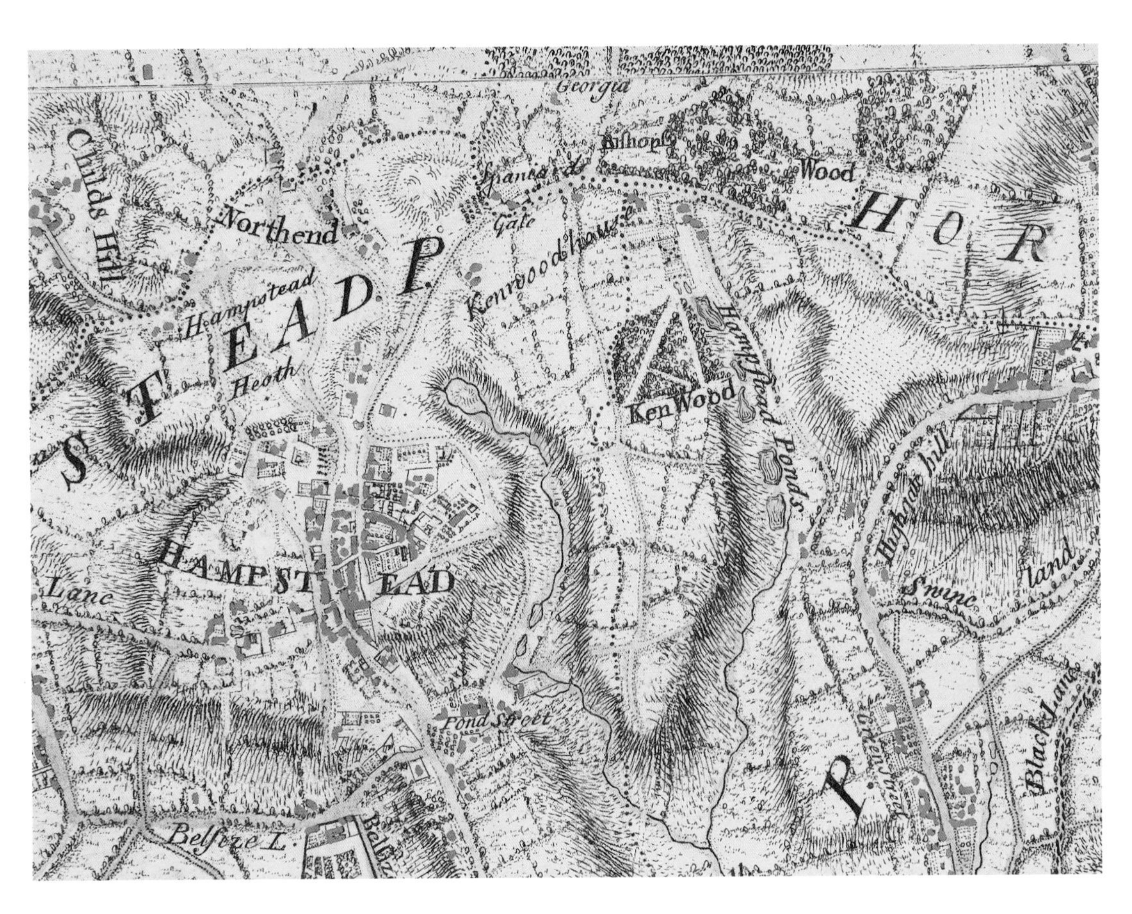

Left: *Detail of Kenwood from a map of London by John Rocque, 1745*
Below left: *The rhododendron garden at Kenwood, added by the 4th Earl of Mansfield in the 19th century*
Below: *Detail from a portrait of John Stuart, 3rd Earl of Bute, owner of Kenwood, by Sir Joshua Reynolds, 1773*

STABLE BLOCK

The stable block was designed by Saunders as part of numerous changes in the 1790s. The Mansfield family would have ridden round the estate on horseback and by carriage, though the area was not very safe. In 1779, the local press reported that the 1st Earl's groom, George Walker, was attacked on Hampstead Lane in an unsuccessful robbery.

Six bedrooms above the stables were listed in the inventory of 1796, as well as harness rooms, a boiling room and a saddle room. The 3rd Earl and his family often travelled between their Scottish property, Scone Palace, by coach to Dundee, boat to London Wharf, then coach to Kenwood. Lord and Lady Mansfield had their own coachman and groom, with other stable staff including under-coachmen, postilions (coach drivers) and stable hands. In 1817, works by the architect Atkinson included repairs to the stalls, coach house and stable doors, as well as '4 new posts to leaping bar' (for a horse jump). In 1854, stable expenditure of £355 5s. 8*d*. included salaries for the footmen, G Fogden and J Dorricott; the groom, D Gould; the coachman, J Birch; and the page, M Hodge.

'...at a very great expense, I repair'd a great house my father had within five miles of Lond[on]; in a situation that yields to none...the whole city within 16 miles of the River appears from every window.'
John Stuart, 3rd Earl of Bute, owner of Kenwood, writing in 1751

Right: Two-piece Reclining Figure No. 5 *by Henry Moore, 1963–4, in the west meadow, on loan from the Tate Gallery, London*
Below right: Flamme *by Eugène Dodeigne, 1983, next to a path below the dairy*
Bottom right: Monolith Empyrean *by Barbara Hepworth, 1953, in the flower garden*

Sculpture at Kenwood

Kenwood is home to three outstanding modern sculptures by Barbara Hepworth (1903–75), Henry Moore (1898–1986) and Eugène Dodeigne (b.1923).

Hepworth and Moore both lived in Hampstead in the 1930s, so it is likely that they would have known Kenwood's collection and estate.

Barbara Hepworth

Barbara Hepworth was one of Britain's leading sculptors and her limestone sculpture entitled *Monolith Empyrean* of 1953 was commissioned by the London County Council and once stood on the South Bank in London. It is now situated in the flower garden between two beds of rhododendrons, near a magnificent handkerchief tree. Its abstract form with an oval head evokes the shape of a person.

Henry Moore

Henry Moore's double-life-sized bronze sculpture, entitled *Two-piece Reclining Figure No. 5* of 1963–4, resembles a recumbent figure, with the round upright representing the head and body and the other part forming the legs. It is on loan from the Tate Gallery and has been displayed on the west pasture ground since 1982. Its position in the landscape seems appropriate for Moore's work, as he liked to situate his sculptures in public spaces for everyone to enjoy.

Eugène Dodeigne

A carved stone sculpture entitled *Flamme*, dating from 1983, can be seen by the path below the dairy. Its rough-hewn form, made of grey granite, looks like a figure, with head and arms. This is a rare example of the work of the Belgian-born, French sculptor Eugène Dodeigne on display in public in Britain.

Left: *The dairy at Kenwood, which was built between 1794 and 1795*
Above: *The tea room inside the dairy, where the ladies of Kenwood had tea. The woodwork has been painted to reflect the appearance of the room in 1834*
Below left: Three Cattle *by Julius Caesar Ibbetson, 1797. The long-horned cattle are grazing on the estate, in front of the newly built dairy*
Bottom left: *The interior of the ice house, an underground brick structure, which held ice brought up from the lakes, for use by the kitchen*

DAIRY AND ICE HOUSE

The dairy consists of three pavilions built between 1794 and 1795 on a hill to the west of the house. It was commissioned by the 2nd Earl and Countess of Mansfield and was probably designed by the architect George Saunders. The buildings include a working dairy with the original marble trays, where cream was allowed to settle at the start of the butter-making process. There is an octagonal tea room, where Lady Mansfield and her guests enjoyed the butter and cream made with milk from cows on the estate. It was fitted out between 1800 and 1801, and paint analysis revealed that it was brightly painted in 1834, with the ornate joinery decorated in vivid shades of blue, black and yellow against a cream ground.

The central building was the dairymaid's quarters. It was remodelled in the 1950s for residential use and now contains offices, as well as being home to a roost of pipistrelle bats. The exterior is covered in lime render with a slate roof.

The ice house was originally one of three on the estate. It consists of an underground egg-shaped chamber made of brick, with straw placed at the base of the chamber to absorb melted water and help drainage. Ice was used for making desserts, sculptures and for packing wine coolers, as well as for keeping food cold when serving. During the winter, carts were hired to transport ice from Kenwood's lake. We know from the household accounts in 1803 that the men were bought beer, bread and cheese from the nearby Spaniards Inn after this tiring work.

Ramsey

Left: A reconstruction drawing of Kenwood in 1756, showing the house before it was remodelled by Robert Adam
Below left: A drawing of Kenwood House by the artist Mary Delany, 1756, showing the original two-storey house
Below: Detail from a portrait of Lady Mary Wortley Montagu, attributed to Jean Baptiste Vanmour, c.1717

Facing page: Detail of Heath House, Hampstead by T Ramsey, 1755, showing people promenading, with Kenwood in the distance

History of the House and Grounds

THE EARLY HOUSE

The first house on the site at Kenwood was probably a brick structure built by John Bill, King James I's printer, who bought the estate in 1616. His son and grandson owned it until 1690, when it was sold to Brook Bridges for £3,400. The house was already substantial, as 24 hearths were recorded in the 1665 hearth tax assessment. It was significantly modified in about 1700, possibly by Brook Bridges's son, William, who owned Kenwood from 1694 to 1705. The new house was a two-storey red-brick building with stone quoins, large sash windows, a hipped roof and a projecting central section topped with a triangular pediment on the south front.

Kenwood changed hands several times in the early 18th century, until it was acquired in 1746 by the Scottish aristocrat John Stuart, 3rd Earl of Bute, George III's future First Minister. Bute's interest in plants probably led to the addition of the orangery on the south-west corner of the site and encouraged him to introduce a number of new plant species to Kenwood.

The footprint of this house first appears on John Rocque's plan of Westminster and Southwark of 1745 and the first views of Kenwood date from a decade later. Mary Delany's drawing of 17 June 1756 shows it as a 'double pile' house, painted white, with a steeply pitched roof and dormer windows.

'I very well remember Caenwood House… I do not question Lord Bute's good taste in the improvements round it, or yours in the choice of furniture.'
Lady Mary Wortley Montagu, mother-in-law of the 3rd Earl of Bute, writing to her daughter in 1749

Right: A reconstruction drawing of Kenwood in 1779, showing the house after it had been remodelled by Robert Adam
Below: Detail from an 18th-century portrait of Lady Mansfield by David Martin
Below right: Portrait of William Murray by Jean Baptiste van Loo, c.1738, before he bought Kenwood in 1754 and became Lord Chief Justice and Lord Mansfield in 1756

'Kenwood is now in great beauty. Your uncle is passionately fond of it. We go thither every Saturday and return on Mondays but I live in hopes we shall now soon go thither to fix for the Summer.'
Lady Mansfield, writing to her nephew in May 1757

KENWOOD, HOME OF THE 1ST EARL OF MANSFIELD

William Murray (1705–93), a distinguished lawyer, acquired Kenwood in 1754 for £4,000 from fellow Scot John Stuart, 3rd Earl of Bute. Soon afterwards, in 1756, he was promoted to Lord Chief Justice, becoming Lord Mansfield, Baron of Mansfield, in the same year. He and his wife, Lady Elizabeth (Betty), née Finch (1704–84), lived in Bloomsbury Square during the week, retiring at the weekends to Kenwood, which at the time was outside London, between the villages of Hampstead and Highgate. Kenwood was a place for relaxation and entertainment – bigger than a town house but smaller than a country seat.

The couple were childless, but from about 1766 they agreed to accommodate Anne Murray, sister of Lord Mansfield's nephew and heir, Lord Stormont; and two great-nieces: Stormont's daughter, Elizabeth Murray, whose mother had died, and Dido Elizabeth Belle, the illegitimate daughter of another of Lord Mansfield's nephews, the naval officer John Lindsay. The decision to house the girls at Kenwood, combined with Mansfield's increasing status and wealth as Lord Chief Justice, encouraged him to remodel the house in 1767 with the help of the architect Robert Adam and his brother James.

Robert Adam first met Lord Mansfield in 1758 through Lord Hopetoun and he visited Kenwood several times in 1760. Adam was an obvious choice of architect, as he was well known to Kenwood's former owner Lord Bute, and in popular demand, working on several similar remodellings of older houses, including Syon House and Osterley Park House, Middlesex.

Lord Mansfield's brief limited Adam to working within the constraints of the existing 18th-century house. He was to improve its external appearance; add a new 'Great Room' for entertaining; accommodate a library; and modernize the existing interiors. The work took place over a period of 15 years, finishing in about 1779.

Contemporaries were impressed by the changes. *The Ambulator, or the Stranger's Companion in a Tour Round London* (1782) described it as, 'Cane Wood, the superb villa of the Earl of Mansfield...The house is magnificent, and the garden front, which is very extensive, is much admired.' Visitors were curious about the building work and were interested in the famous owner of the house. Lord Mansfield and his wife were also engaging hosts. According to the dramatist and civil servant Richard Cumberland (1732–1811), Lord Mansfield possessed 'the happy and engaging art...of putting the company present in good humour with themselves: I am convinced they naturally liked him the more for his seeming to like them so well'.

Lord Mansfield entertained a wide range of people at Kenwood outside his family circle, including statesmen, politicians, lawyers, artists and even royalty. King George III and Queen Charlotte were reputed to have been frequent visitors to Kenwood, according to the artist William Birch (1755–1834).

Above: *Portrait of Lady Elizabeth and her sister Lady Henrietta Finch, c.1730–1, by Charles Jervas. Lady Elizabeth's black dress, painted over a lighter coloured fabric, may show her in mourning for her father, who died in January 1730. She married William Murray in 1738*

Left: *Engraving of the north front of Caen Wood in 1788, showing Hampstead Lane in front of the house before it was re-routed*

Right: View of the library at Kenwood. Lord Mansfield was a bibliophile and owned a fine collection of books

Below right: Robert Adam's design for the north wall of the library of 1767, showing different options for sofa designs

ROBERT ADAM AT KENWOOD

The first evidence for the involvement of the Adam brothers at Kenwood dates from 1764, in a series of drawings, now in Sir John Soane's Museum. Four designs by James Adam show his unexecuted proposals for a new combined 'Great Room' and library, including ideas for the decoration of the ceiling, with an unexecuted design by Robert for the south front. Work began in 1767, when drawings were completed by Robert for the ground plan and south front of the house. These designs show how he worked with the layout of the existing house, but developed a scheme to add the 'Great Room' or library onto the south-east elevation to counterbalance the existing orangery on the west, using the new antechamber and staircase hall to link the old and new parts of the building.

From 1767 to 1779, Adam progressed through the interiors at Kenwood, updating and redecorating them with his team of workmen, which included the plasterer Joseph Rose, the woodworker John Minshull, the painter–decorator George Steuart and the decorative artist Antonio Zucchi. Adam redecorated all the rooms on the ground floor. The entrance hall was refitted in 1773 and the staircase, the new antechamber and the new library were built between 1767 and 1769. The south front rooms, which were used as more intimate social spaces for the family, including a drawing room, parlour and dressing rooms for Lord and Lady Mansfield, were also redecorated by Adam.

Adam also worked on the first floor in Lord and Lady Mansfield's bedchambers, inserting a chinoiserie chimney piece in the upper hall. The extent of his alterations on this floor is difficult to assess, however, as no floor plan survives. Adam added the second floor, or attic storey, to the house with minimal disruption, by building the new walls and adding the roof before removing the old one so that the house could remain habitable.

Adam also sought to unify the appearance of the outside of the house with elegant stuccoed elevations. He designed the new entrance on the

Mansfield and the Campaign for the Abolition of Slavery

By the 1780s Lord Mansfield was considered to be one of England's greatest judges.

In 1754 he became Attorney-General, followed in 1756 by his appointment as Chief Justice of the Court of King's Bench, when he was also created Lord Mansfield, Baron of Mansfield, becoming Earl Mansfield in 1776.

Contemporaries regarded him as 'that excellent useful Judge but mischievous politician' and in 1775 Samuel Curwen, a loyalist American merchant from Salem, commented on his 'piercing eyes [which] denote a penetration and comprehension peculiarly his'.

Among his important legal cases was the trial of John Wilkes MP for seditious libel. He also opposed the demands of the American colonies for independence. As a legislator he was most interested in social issues, working on a Bill in 1777 to protect young women from seduction. He is famously associated with judgments that contributed to the abolition of slavery, most notably in the 'Somerset' case of 1772. James Somerset, a former slave, was brought before the court by a writ of *habeas corpus* obtained by the abolitionist Granville Sharp. Somerset had been imprisoned by his former master and was being shipped out to be sold in Jamaica.

Lord Mansfield's verdict was that a master could not by force dispatch a slave out of the country, declaring that, 'slavery…is so odious that it must be construed strictly'. An American visitor to Kenwood in 1779 described Lord Mansfield's explanation that the judgment 'went no further than to determine the Master had no right to compel the slave to go into a foreign country'. But Mansfield's personal interest in the rights of slaves was apparently evident in his pleasure at having released 'two Blacks from slavery, since his being Chief Justice'.

__Left:__ William Murray, 1st Earl of Mansfield, by David Martin, 1775. As Lord Chief Justice, the Earl of Mansfield is remembered most often for his rulings in favour of black plaintiffs

__Above:__ Pendant distributed to supporters of the abolition of slavery, with the motto 'Am I not a man and a brother?' made by Josiah Wedgwood and Sons in about 1787

__Below:__ Detail from a slave sale in New Orleans, engraved by J M Starling in 1842

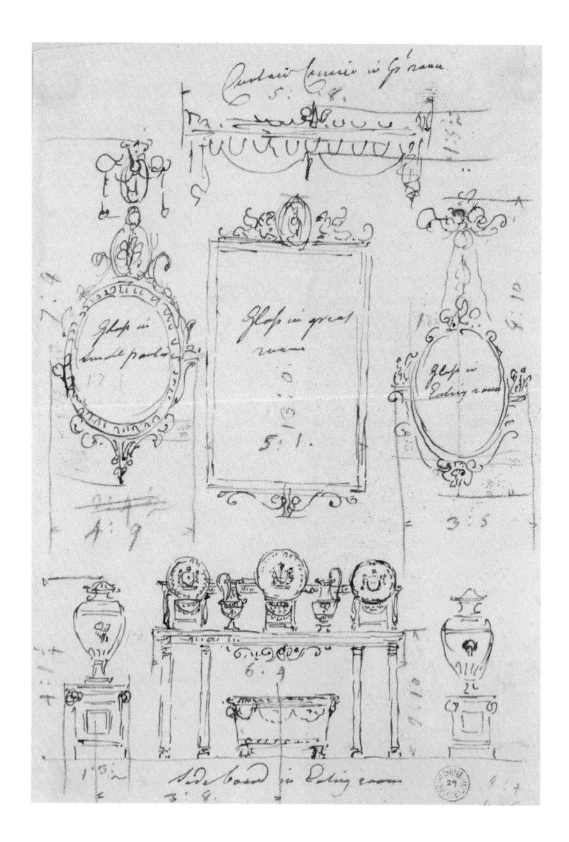

Right: *Preliminary drawing for various items of furniture for Kenwood by Robert Adam, 1774, including a sideboard table for the entrance hall*
Below right: *The mob burning, plundering and freeing prisoners of Newgate Prison in 1780, during the Gordon Riots, when Kenwood narrowly escaped damage*

north front in 1764, creating a giant, full-height, pedimented portico. The south front elevation was also originally designed in 1764, but was redesigned in 1768 to accommodate the attic storey bedrooms. Painted in a stone colour, the façade was covered in 'Liardet' oil cement – Adam's newly invented patented stucco – in which his architectural firm had a financial interest, but which proved problematic, as it had to be repaired constantly.

Adam provided Lord Mansfield with remarkable new interiors at Kenwood, as well as some designs for furniture, most notably for the library. The work was largely completed by 1774, when it was included in the second part of *The Works in Architecture of Robert and James Adam*, published in 1774. Adam explained how Lord Mansfield 'gave full scope to my ideas' and he described the house's position: 'to the north east, and east of the house and terrace, the mountainous villages of Highgate and Hampstead form delightful objects. The whole scene is amazingly gay, magnificent, beautiful and picturesque. The hill and dale are finely diversified, nor is it easy to imagine a situation more striking without, or more agreeably retired and peaceful within'. Only two other houses – the Duke of Northumberland's Syon House and Lord Bute's Luton Hoo – were published in *The Works in Architecture of Robert and James Adam*, testimony to Adam's pride in his achievement at Kenwood.

The Gordon Riots

In June 1780, Kenwood had a narrow escape during the Gordon Riots, a series of violent anti-Catholic protests, when the house was almost burnt to the ground. Kenwood was a target, as it was the home of the Lord Chief Justice.

In 1778, an Act of Parliament had been passed to allow Catholics greater rights, excusing them from swearing the oath of allegiance to the Crown on joining the army (with its implicit recognition of the Church of England). Lord George Gordon subsequently attempted to repeal this Act and on 2 June 1780, he led a crowd of 60,000 men to the House of Commons to present a petition stating that the legislation encouraged 'popery' and was a threat to the Church of England. Anti-Catholic rioting erupted, fuelled by fears of the rise of papism and the return of absolute monarchical rule.

As a Scot, Lord Mansfield was suspected of being a Jacobite and the mob attacked his house in Bloomsbury Square, destroying his important legal library. As Lord Chief Justice, he was ultimately responsible for the Catholic Relief Act of 1778 and his Catholic sympathies and acknowledged belief in religious tolerance made him an obvious target. The protesters then marched towards Kenwood but they were intercepted en route by Lord Mansfield's nephew and heir, Lord Stormont, who dispatched a detachment of light horse to stop them getting close to the house and waylaid them with refreshments at the nearby Spaniards Inn.

The Story of Dido Belle

Dido Elizabeth Belle (1761–1804) was the illegitimate daughter of Sir John Lindsay (1737–88) and a black slave called Maria (c.1746– after 1781).

Sir John was a Royal Navy officer and a nephew of the 1st Earl of Mansfield, and Maria was an enslaved African. The circumstances of Dido's birth in 1761 are unconfirmed, but her father captained the frigate *Trent* at this time so she was probably conceived during his voyage to Jamaica. By 1766, when Dido was baptized on 20 November at St George's church, near the Mansfields' Bloomsbury home, Maria was in London with her husband, Mr Bell. Dido's mother was still living in London in 1774 and recorded as 'free' before moving to Pensacola, Florida, where Sir John gave her a piece of land.

Dido grew up in Lord Mansfield's household with her cousin, Lady Elizabeth Murray (c.1763–1823), who is shown beside her at Kenwood in this rare 18th-century depiction of a mixed-race woman beside her white relation.

We know that Sir John was a favourite nephew of Lord Mansfield but we do not know how often he visited Kenwood or if he had any kind of relationship with Dido. We know that Dido was educated, literate and oversaw the dairy, a duty often given to the lady of the house. By comparing her annual allowance and the money she was given for her birthday and for Christmas, it is clear that her status within the household was higher than that of a servant but generally below that of the rest of the family.

Her father's obituary in *The London Chronicle* in 1788 states that her 'amiable disposition and accomplishments have gained her the highest respect from all his Lordship's relations and visitants'. Lord Mansfield was fond of Dido and in his will he left her £500 and an annual allowance of £100 and confirmed her free status. In 1793 Dido married John Davinier, a steward (a senior servant), and they are known to have had three sons and lived in Pimlico, London, until her death, aged 43.

fm Sep 25 to Octo 1 . . . . H . . 13 1 3½
Dido's ¼ allowance due 4 5
Jane Jones's Years Wages due 8 8

Above: *Thomas Hutchinson, Governor of Massachusetts, by Edward Truman, 1741. Hutchinson visited Kenwood in 1779 and wrote, 'A Black came in after dinner and sat with the ladies, and after coffee, walked with the company in the gardens, one of the young ladies having her arm within the other.'*

Left: *Cousins Dido Belle and Lady Elizabeth Murray portrayed at Kenwood, previously attributed to Johann Zoffany, c.1775–85*

Below left: *Detail from Anne Murray's account book, covering the period from 1785 to 1793, listing Dido's quarterly allowance of £5 in 1785*

Above: *Miniature portrait of David Murray, 7th Viscount Stormont, later 2nd Earl of Mansfield, after a portrait by Pompeo Batoni, 1768*
Below right: *A reconstruction drawing of Kenwood in 1797, after the changes to the landscape made by Repton*

KENWOOD UNDER THE 2ND EARL OF MANSFIELD

In 1793, David Murray (1727–96), 7th Viscount Stormont, and former ambassador to Vienna and Paris, succeeded his uncle to become 2nd Earl of Mansfield. He knew Kenwood well, as his sister Anne and daughter Elizabeth had been living with his uncle in London since 1766, and he had visited many times, including during his honeymoon in 1776 with his second wife, Louisa. He had clearly already decided to enlarge Kenwood to accommodate his family, claiming that 'offices were absolutely necessary, and as Lord Mansfield had so frequently recommended to me the embellishment of Kenwood I resolved that they should be upon a handsome plan'.

Building Work at Kenwood

The first thing he did from May 1793 was to construct the new road, which diverted Hampstead Lane away from the end of the original forecourt, to provide the house with more privacy. Building work on the service wing (then called the 'office wing') began in June 1793 and was noted in November by the artist and diarist Joseph Farington, who observed the 'considerable, and in respect of architectural effect, strange additions to the late Lord Mansfield's house at Caenwood'.

These changes included the addition of two brick wings to the east and west of the north front, which contained a dining room and the Music Room, as well as a service wing, with kitchens, bedrooms, brewhouse and laundry, new gate lodges, a new farm, dairy and stables. The Music Room, which also served as a drawing room, was decorated by the artist Julius Caesar Ibbetson (1759–1817) and faced onto the garden. The 2nd Earl decided to leave the wings unornamented and bare of stucco, to distinguish them from the earlier house, remodelled by Adam.

The 2nd Earl originally chose the architect Robert Nasmith, but his death in August 1793 led to the appointment of George Saunders, who had been acting as surveyor at Kenwood and was also employed at Scone Palace. From 1793 to 1797 Saunders supervised the substantial building works at Kenwood.

REPTON AND THE CREATION OF THE LANDSCAPE AT KENWOOD

The 2nd Earl of Mansfield also enhanced the grounds at Kenwood, employing the celebrated landscape architect Humphry Repton, who spent several days here from 8 May 1793. Repton's approach was to evaluate the topography and potential of a site and to present his proposals

Left: *Two drawings of the north front of Kenwood, from Repton's 'Red Book', showing the 'before' and 'after' views of his proposed changes*
Below: *Portrait miniature of Humphry Repton by John Downman, c.1790*

in drawings and words in a 'Red Book', which included 'before' and 'after' sketches.

Repton's 'Red Book' for Kenwood dates from May to July 1793 and illustrates his suggestion to transform the house into a country mansion and re-landscape the grounds. He asked the 2nd Earl to introduce him as soon as possible to the architect, Nasmith, so that 'our operations both within and without doors may harmonise with each other'. Following Nasmith's death in August 1793, Repton tried unsuccessfully to persuade his patron to employ the architect William Wilkins (d.1815) as his successor.

Repton's drawings for the north front show the construction of two new wings on either side of Adam's portico and the masking of unsympathetic buildings with planting. As early as May 1793, the 2nd Earl had already begun building works to relocate Hampstead Lane, which was pushed back to provide a better approach to the house. Repton's drawing shows the creation of two sweeping drives to the east and west of the north forecourt, by which visitors would arrive at the house. Service buildings, including the kitchens and offices, were to be rebuilt and located in a 'deep valley' to the east, where they could be hidden behind banks and shrubbery. He recommended that a dining room be added to the east of the portico with a beautiful view to the north. To the west of the house, where the kitchen garden had originally been located, Repton proposed a new lawn and flower beds. Adam's much admired south front – described by Repton as 'that magnificent terrace which is doubtless one of the first ornaments of Kenwood' – was probably least affected, although Repton's proposals showed plans to open up the narrow, tree-edged south lawn and to construct new carriage drives. His ambitions to 'acquire a view totally unlike any other at Kenwood, and indeed superior in splendour to most others in the kingdom' would have flattered his patron.

'A superb and elegant Mansion, surrounded by a sufficient extent of landed property, to give all the importance, convenience and even privacy, of many situations in more distant parts of the Kingdom.'
Humphry Repton, writing in 1793

'Lord Stormont and I are now so well acquainted that I wonder how I ever could be afraid of him.' Louisa Cathcart, talking to her sister in 1776

LOUISA CATHCART, COUNTESS OF MANSFIELD, AND THE CREATION OF THE DAIRY

Louisa, daughter of the 9th Baron Cathcart, married Lord Stormont, the future 2nd Earl of Mansfield, in 1776, as his second wife. Louisa was 31 years his junior. She was interested in agricultural improvements and was responsible for the dairies at Scone Palace and Kenwood.

Dido Belle had been in charge of the dairy at Kenwood in 1779 but the current dairy is a later building, designed for Louisa by George Saunders between 1794 and 1795. It comprised three buildings, including a small octagonal tea room, a 'Dairy House' and a 'Scullery', with an ice house below. Workmen's accounts refer to the marble milk pan and black marble basin. It was a working dairy, and was later taken over by the Express Dairy Company Ltd as one of its branches in the 1890s.

The popularity of dairies in the 18th century was influenced by the French queen Marie Antoinette's dairy at the château of Versailles, which Louisa may have visited with her husband during his ambassadorial posting to Paris.

As mistress of the house, Louisa presided over the dairy, which supplied the household with butter, milk, cream and cheese. A contemporary noted the competition between Lady Mansfield and her neighbour, Lady Southampton of Fitzroy Farm, both of whom were 'admirable dairy-women…[who] were so jealous of each other's fame, that they have…been very near to a serious falling out, on the dispute [over] which of them could make the greatest quantity of butter from such a number of cows'. Louisa also probably commissioned the paintings of Kenwood's longhorn cows.

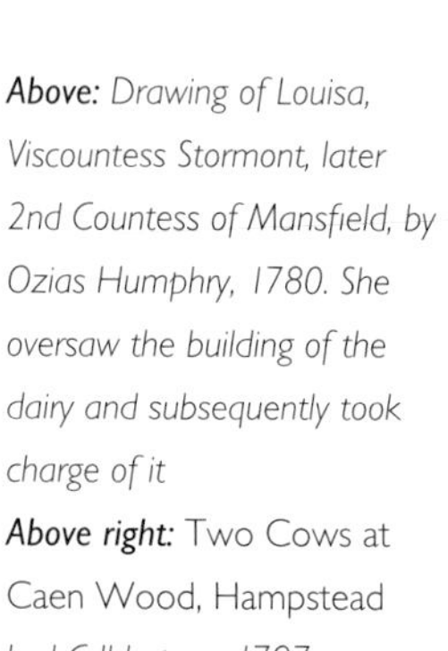

Above: *Drawing of Louisa, Viscountess Stormont, later 2nd Countess of Mansfield, by Ozias Humphry, 1780. She oversaw the building of the dairy and subsequently took charge of it*
Above right: Two Cows at Caen Wood, Hampstead *by J C Ibbetson, 1797, a painting probably commissioned by Louisa, 2nd Countess of Mansfield*
Right: *The interior of the dairy with its surviving marble floor, milk basins, window shutters and wall tiles*

KENWOOD UNDER THE LATER EARLS

In 1796, David William Murray, 3rd Earl of Mansfield (1777–1840), inherited Kenwood at the age of 19, marrying Frederica Markham, daughter of the Archbishop of York, the following year. Although his father had bequeathed him £20,000 to complete the building work at Kenwood, he preferred to live at Scone Palace in Perth. By 1815, however, a number of structural problems at Kenwood forced the 3rd Earl to appoint his architect from Scone, William Atkinson, to carry out repairs and alterations. Dry rot had attacked the building and Atkinson was forced to strengthen the Great Stairs and the Deal Stairs and rebuild the north wall of the hall. Works included the addition of second-floor rooms to the service wing; the installation of bookcases in the mirrored niches in the library; and extensive redecoration that included imitation red porphyry for the columns and pilasters in the antechamber.

During the 19th century, the Mansfields largely discouraged general public access to the house and grounds, although they celebrated the visit of King William IV in 1835 with a grand luncheon in the dining room (see below).

William David Murray (1806–98), who became 4th Earl in 1840, continued to favour the family's Scottish seat at Scone, although he did spend three months a year at Kenwood. On his instruction, large French windows were inserted in the Music Room and a veranda was built to give access to the west lawn. The 4th Earl was also instrumental in helping to preserve Hampstead Heath from developers, through his purchase of adjacent land to the east and the sale of Parliament Hill Fields to the Metropolitan Board of Works (later London County Council) in 1889, which greatly increased the public area of the Heath.

Left: William David Murray, 4th Earl of Mansfield, who spent three months a year at Kenwood
Above: Alan David Murray, 6th Earl of Mansfield, who inherited Kenwood from his brother and put the first electric lights in the house in 1907
Below left: Portrait of Fanny Gascoyne-Cecil by Sir Thomas Lawrence, 1829

By contrast, William David Murray, the 'bachelor' 5th Earl (1860–1906), who inherited in 1898, entertained lavishly at Kenwood. He also actively campaigned to preserve Hampstead Heath from development. His sudden death from pneumonia in 1906 led to the succession of his brother, Alan David Murray (1864–1935) as 6th Earl. He introduced electricity to the house in 1907, but preferred to live at Scone Palace and he rented Kenwood out to tenants.

King William IV's Visit, 1835

'The breakfast at Ken Wood. The road was crowded with people all the way anxious to see the King.
A triumphal arch was erected on Hampstead Heath, and in most of the houses by the side of the road there were preparations for illuminations. I heard the King was extremely well received by the crowd, and the Duke still more so… It was a beautiful day. The grounds are excessively pretty, and if there had been enough to eat, it would have been perfect…The King and Queen and all the Royalties seemed extremely well-pleased: the King in particular trotted about with Lord M. in the most active manner, and made innumerable speeches!'
Fanny Gascoyne-Cecil, Marchioness of Salisbury (1802–39), writing in her diary, Thursday 23 July 1835

Above: *Grand Duke Michael with his dog and children: Nada, Zia and Michael, c.1903*
Right: *Grand Duke Michael playing tennis at Kenwood, on the lawn in front of the orangery, c.1915*
Below: *The wedding of Grand Duke Michael's daughter Nada to Prince George of Battenburg at Kenwood in 1916*

KENWOOD'S TENANTS: THE GRAND DUKE MICHAEL AND NANCY LEEDS

In September 1909, the *Daily Telegraph* reported that 'His Imperial Highness the Grand Duke Michael of Russia planned to lease Kenwood from the Mansfield family and move there with his wife, the Countess Torby…as soon as certain alterations…including the laying out of a new golf course have been carried out'. The Grand Duke Michael Michaelovitch (1861–1929), grandson of Tsar Nicholas I and second cousin to the last Tsar Nicholas II, was exiled from Russia in 1891, following his morganatic marriage to Sophie, a great-granddaughter of the Russian poet Pushkin. The couple signed a 21-year lease on the furnished property at a cost of £2,200 a year.

The family moved into Kenwood in 1910 with their three children: Michael, Nadejda (Nada) and Anastasia (Zia). They participated in local life, with the Grand Duke becoming President of the Hampstead General Hospital. They played tennis on the lawn and entertained lavishly, playing host to King George V and Queen Mary, who attended a ball here in June 1914.

A highlight of their stay at Kenwood was the marriage in 1916 of their eldest daughter, Nada, to Prince George of Battenberg, uncle of the Duke of Edinburgh, with guests including King George V and Queen Mary. Photographs of Kenwood taken for *Country Life* in 1913 give the best evidence of

the interior furnishings before the original furniture was sold in 1922.

During the First World War, Kenwood was offered for use as a hospital, and in November 1915 the Royal Naval Anti-Aircraft Mobile Brigade was established in the stable block and the stables were used as barracks. In 1917, however, the Grand Duke and his family were forced to sublet Kenwood, following the Russian Revolution and the murder of his cousin the Tsar. As Grand Duke Michael wrote to Arthur Balfour, Foreign Secretary, 'His Imperial Highness is entirely without means for his living requirements and has no capital or money here.' Queen Mary lent the family £10,000 and wrote sympathetically to Sophie: 'What a grief leaving your beautiful Ken Wood which you made so charming and comfortable, but I hope you will find a small house to your liking and that before long things may improve all round. Poor poor Russia!' The only surviving signs of Kenwood's Russian occupancy are two memorial tablets to the Grand Duke's dogs, 'Bill' and 'Mac', who were buried in North Wood (see page 47, opposite).

In 1917, Mrs Nancy Leeds (née Stewart), widow of American tin millionaire William B Leeds, took on the lease of Kenwood. She did not live here long, moving out in 1920 when she married Prince Christopher of Greece.

Left: The Royal Naval Anti-Aircraft Mobile Brigade at Kenwood between 1915 and 1916

Below left: Grand Duke Michael of Russia, Admiral Sir Percy Scott and Commander Rawlinson at Kenwood, 1915–16

Below: Grand Duke Michael lived at Kenwood during the First World War. The only record of his time at Kenwood is the gravestone of his dog, Mac, who died in 1915 and is buried in North Wood. The inscription reads: 'To my old and faithful friend Mac'

World Wars at Kenwood

The Royal Naval Anti-Aircraft Mobile Brigade was stationed in the stable block at Kenwood from November 1915 until August 1916.

Commander A Rawlinson led the unit and later published his memoirs, which included anecdotes about Kenwood. Grand Duke Michael of Russia was living at Kenwood at the time. Commander Rawlinson describes breakfasts with the duke and his family at the house, noting the contrast with the extreme poverty and despair in the East End of London. Officers and men slept in hammocks in the stable barracks. They used searchlights and high-angle anti-aircraft cannons mounted on Lancia lorries during night-patrol, awaiting the bomb attacks by Zeppelins over London.

Second World War

Kenwood closed its doors shortly before the Second World War. Some of the paintings were taken off site to the National Portrait Gallery's store but the majority of the collection remained with the house manager, Captain Roberts-Wray, and the secretary and housekeeper, Miss Mary Tibbs. In 1946 some Old Masters were recorded stacked on their sides in one of the roof spaces.

During the Second World War, Kenwood was used as overflow accommodation for the RAF Intelligence School, with servicemen attending lectures in the dining room, and living on the first floor with the Breakfast Room as the dining hall and Lord Mansfield's Dressing Room serving as the pool room.

THE 1ST EARL OF IVEAGH AND THE RESCUE OF KENWOOD

While Kenwood was let, the 6th Earl of Mansfield had entered into negotiations to sell the house and its estate in 1914 and there were protracted discussions about its future. A group of local residents managed to negotiate with Lord Mansfield, who offered the house and 220 acres for £550,000. Discussions were put on hold during the First World War but resumed in 1918 with the newly formed Kenwood Preservation Council, led by the rich industrialist Sir Arthur Crosfield. In 1920, the Kenwood Preservation Council was offered an option on the land for £340,000 but could only raise £85,000 before the option expired on 1 December 1921. The council had to settle for buying 100 acres in the south of the estate, including the meadow lands and Grand Duke Michael's golf course, for £135,000 in December 1922.

Plans to develop the land around the house continued, with 33 spacious villas each with one- and two-acre gardens marked out in plots. Meanwhile, Lord Mansfield sold the house contents at an auction run by the local Highgate firm of C B King. Almost all of the Kenwood items collected by the 1st, 2nd and 3rd Earls, apart from some items removed by the family to Scone Palace, were dispersed at this four-day sale in November 1922.

In 1924 the Kenwood Preservation Council vested land, including the ponds to the south of the house and a further 32 acres purchased that year, in the London County Council. Kenwood's grounds were officially opened as a public park by King George V on 18 July 1925.

During this time, Edward Cecil Guinness, 1st Earl of Iveagh, was a Hampstead resident, with two houses not far from Kenwood: Heath House and Heathlands. On 31 December 1924 he took a ten-year lease on Kenwood and soon afterwards his family trust purchased the house and the remaining 74 acres of land for a much reduced price of £107,900. The Iveagh family had a strong history of philanthropy. Lord Iveagh intended to give the house and estate to the nation, together with a selection of his paintings, after a decade.

Above: *Sir Arthur Crosfield, who was instrumental in saving Kenwood from property developers, photographed by Bassano Ltd, 1920*
Right: *Poster by 'Poy' (Percy Fearon) for the Kenwood Preservation Council, 1921*
Below right: *King George V, filmed by British Pathé News at Kenwood on 18 July 1925, when the estate was opened to the public*
Below left: *Rupert Cecil Guinness, 2nd Earl of Iveagh, who oversaw the opening of Kenwood to the public, photographed by Howard Coster, 1931*

'My father during the First World War spent a good deal of time up here at Heath House in Hampstead, and he was very much impressed and delighted with Hampstead and the country around. At that time the place was for sale, and his idea was to make a wonderful, I think a most wonderful idea, and that was to give some of his best pictures that they should be in the proper surroundings… where everybody could appreciate them.'
Rupert Guinness, 2nd Earl of Iveagh, talking about his father in 1950

Left: *Portrait of Edward Cecil Guinness, 1st Earl of Iveagh, by H M Paget, after A S Cope, painted after 1912*

Right: *Among Lord Iveagh's collection are portraits of children, such as this charming painting of Miss Murray by Sir Thomas Lawrence, 1824–26*
Below: *Portrait of a Young Girl by Friedrich von Amerling, c.1835–40. This painting was smuggled out of Vienna by Ms Heller-Binder in March 1939. Her father was a leading art dealer in Vienna and she even remembered Hitler visiting her father's gallery. She donated the painting to Kenwood in 1973, so adding to the house's collection of paintings*

THE IVEAGH BEQUEST

Lord Iveagh died in 1927 before he could supervise the acquisition and arrangement of furnishings for the rooms. He had chosen Kenwood as a suitable home for displaying highlights from his world-class painting collection, for the public to see and enjoy. After his death, there were six administrative trustees: four nominees of the Guinness family, including his son, the 2nd Earl; a representative from London County Council; and Charles Holme, the Director of the National Gallery, who assisted with the hanging of the paintings. The trustees decided how best to decorate the interiors and display the pictures and furniture according to Lord Iveagh's wishes. Kenwood opened to the public on Wednesday 18 July 1928.

The Iveagh Bequest Act of 1929, which established Kenwood as an independent museum, clarifies Lord Iveagh's intention that the house and its contents were intended to present 'a fine example of the artistic home of a gentleman of the eighteenth century'. The type of paintings displayed, mostly Old Masters and 18th-century English portraiture, reflect this idea. The spirit of the bequest continues to be upheld today.

The Decorative Arts Collection

The miniature, jewellery and buckle collections, on display in the miniatures room on the mezzanine level, have all been donated by local benefactors.

Draper Gift

Marie Elizabeth Jane Irving Draper (1940–87) was fascinated by 18th-century portraiture and she started to collect miniatures in the 1970s, when she developed glaucoma and tunnel vision, as their small size and portability allowed her to see them clearly. Her mother, Elizabeth Pearce, expanded the original gift of 20 miniatures with funds from her daughter's estate. Today, the Draper Gift contains over 100 portrait miniatures, including examples by Richard Cosway (1742–1821) and George Engleheart (1750–1829).

Hull Grundy Collection

The jewellery was gifted in 1975 by Anne Hull Grundy (1927–84), scholar, antiquarian and collector. Anne fled Germany with her parents in the 1930s and settled in Hampstead, London. She was a self-taught connoisseur who began buying jewellery as a teenager. She often bought out-of-fashion items, such as Victorian brooches, for reasonable sums. She donated thousands of jewels and objets d'art to museums throughout Britain, many to the British Museum, and she handpicked 'a dazzling display of Georgian splendour' specifically for Kenwood.

Lady Maufe Collection

The shoe buckles were gifted in 1971 by Lady Prudence Maufe (1882–1976), a designer, interior decorator and director of Heal's. She began collecting historic shoe buckles in the 1960s and within ten years had formed a remarkable collection of many thousands. Buckles were extremely fashionable for holding shoes in place, before the use of laces.

Left: *Some of the Lady Maufe Collection of shoe buckles, on display at the Wellington Arch in 2012. The collection includes items from many countries, showing a fascinating variety of manufacturing techniques and materials*

Above, from left to right: *Miniature of an unknown woman by Richard Cosway, c.1785–95; a Victorian mosaic brooch with a view of Rome, collected by Anne Hull Grundy; and a miniature of an unknown man by George Engleheart, c.1785–90. The two miniatures are part of the Draper Gift*

Right: The south front of Kenwood, shrouded in early morning mist
Below: Repainting the library in 2013 involved over-painting the gilding and restoring the room to its original colour scheme

KENWOOD IN THE 21ST CENTURY

English Heritage has been responsible for the Iveagh Bequest at Kenwood since 1986, carrying out extensive repairs to the exteriors, transforming the landscape and completing substantial redecoration projects to the interior, while continuing to add items of suitable quality to the collection of furniture and paintings. It has endeavoured to take a historically researched and unified approach to the house and estate, so that visitors can enjoy both the villa and its grounds.

The redecoration of the two north wings in 2000 involved the paintings being hung in more fully furnished interiors. Research and analysis underpin these major projects, most recently exemplified when the house closed for roof repairs, exterior redecoration and the re-presentation of eight rooms on the ground floor. Some of the Adam interiors, most notably the library, have been transformed. The original colour schemes have been reinstated, which more accurately reflect the appearance of the rooms in the time of the 1st Earl of Mansfield.

The tradition of the annual outdoor concerts continues on the estate in the summer, allowing thousands of visitors to see and enjoy the Iveagh Bequest, Kenwood and its beautiful grounds.